WATER CONTAMINATION & SCARCITY

Dr. Abu Adil

RIGI PUBLICATION

Water Contamination & Scarcity

By

Dr. Abu Adil

Originally Published in India

ISBN: 978-93-95773-69-0

Published by RIGI PUBLICATION

777, Street no.9, Krishna Nagar
Khanna-141401 (Punjab), India
Website: www.rigipublication.com
Email: info@rigipublication.com
Phone: +91-9357710014, +91-9465468291

PREFACE

In the heart of the picturesque landscapes and vibrant culture of Jammu lies a silent crisis, one that strikes at the very essence of life itself: water scarcity and contamination. As the author of this book, I embarked on a journey to explore the depths of this critical issue that plagues our region. This exploration was not merely an academic pursuit but a heartfelt endeavor, born out of concern for the well-being of our communities and a profound love for this land I call home.

As I delved into the research, I was confronted with harsh realities. The water sources that have sustained generations are dwindling at an alarming rate, leaving in their wake parched lands and desperate lives. Simultaneously, the contamination of available water, once pure and life-giving, has become a looming threat, casting a shadow over the health and prosperity of the people of Jammu.

About the Author

Dr. Abu Adil

Dr. Abu Adil, a dedicated advocate for sustainable and inclusive development, has devoted his career to researching and understanding the challenges faced by underserved populations in the union territory of Jammu and Kashmir. With his masters degree in social work and MD in prevention and social medicine, Dr. Adil founded and continues to direct a non-government organization that serves tribal and nomadic populations through healthcare, education, economic empowerment, and safe drinking water services.

For the last sixteen years, Dr. Adil has developed a deep connection to this region that spurred his passion to conserve natural resources for the generations to come. He has witnessed the challenges faced by many tribal and nomadic communities due to water contamination and scarcity, which has been the driving force of his research into the complexities of this issue and his resolve to seek viable solutions.

Dr. Adil's research journey has taken him across remote villages and bustling cities in Jammu, where he has engaged with local communities, experts, and policymakers to gain a comprehensive understanding of these water-related challenges. His hands-on approach has allowed him to collect valuable data and stories, illuminating the human side of this crisis.

His tireless efforts have earned him recognition in the field of health care and education. Dr. Adil was honoured with The Mahatma Gandhi Sewa Puraskar Award by the Honourable Lieutenant Governor of Jammu and Kashmir Union Territory and a Nobel Award by the Global Human Rights Organization.

In this book, Dr. Adil combines his extensive research, personal experiences, and unwavering passion to shed light on the pressing concerns of water contamination and scarcity in Jammu. Through meticulous analysis and compelling narratives, he not only highlights the challenges faced by communities, but also proposes practical solutions and calls for collective action.

Dr. Adil's expertise, coupled with his dedication to ensuring access to safe water for vulnerable communities, makes "Water Contamination and Scarcity" a definitive and inspiring read. His work serves as a beacon of hope, encouraging readers to join the fight for clean and accessible water in Jammu and across the globe.

WATER CONTAMINATION & SCARCITY

CONTENT

Introduction to the importance of water and its scarcity in Jammu

Water is a fundamental resource essential for all life forms, making it a cornerstone of human civilization. Its importance cannot be overstated, as it is vital for various purposes such as drinking, sanitation, agriculture, and industrial activities. However, despite its significance, many regions, including Jammu, face the critical issue of water scarcity.

Jammu, a picturesque region in northern India, is blessed with natural beauty, but it grapples with the scarcity of water, especially in recent years. The significance of water in Jammu's context lies not only in sustaining life but also in supporting its agrarian economy. Agriculture, being a primary occupation in the region, heavily relies on water for irrigation. The scarcity of water disrupts agricultural activities, leading to decreased crop yield and economic challenges for farmers.

Furthermore, water scarcity affects the health and well-being of the residents. Insufficient and contaminated water supply can result in waterborne diseases, posing a significant threat to public health. Access to clean and safe drinking water is essential for preventing such diseases and ensuring the overall health of the community.

Water scarcity also has far-reaching environmental implications. Depletion of water sources can lead to soil erosion, loss of biodiversity, and adverse effects on aquatic ecosystems. The delicate balance of nature is disrupted, affecting not only the environment but also the lives of the people who depend on it.

In light of these challenges, it becomes crucial for the community, government, and various stakeholders to address the issue of water scarcity in Jammu. Sustainable water management practices, such as rainwater harvesting, efficient irrigation techniques, and water conservation awareness programs, can play a pivotal role in mitigating the problem. Additionally, the implementation of policies and infrastructure projects aimed at ensuring a consistent supply of clean water is essential.

In conclusion, water is a precious resource that sustains life, supports livelihoods, and maintains ecological balance. Its scarcity in regions like Jammu highlights the urgent need for collective efforts in water conservation and sustainable management. By recognizing the importance of water and taking proactive measures, we can secure a better future for the people of Jammu and preserve this invaluable resource for generations to come.

Historical overview and water management in Jammu

The historical overview of water management in Jammu reveals a fascinating journey of innovation, adaptation, and challenges faced by the region in harnessing and distributing its water resources. Throughout history, water management in Jammu has played a vital role in shaping its agricultural practices, economy, and societal structures.

In ancient times, the region's water management primarily revolved around harnessing the natural water sources like rivers, springs, and lakes. The innovative use of step wells, known as baoli's was prevalent. These baoli's were intricate structures designed to conserve rainwater and provide a reliable source of water for drinking and irrigation. The ingenuity of these structures showcased the early society's understanding of the importance of water conservation.

During the medieval period, various rulers and dynasties, including the Dogras, contributed significantly to water management in Jammu. They constructed a network of canals, reservoirs, and aqueducts to channel water for agricultural purposes. The construction of Ranbir Canal in the late 19th century under the rule of Maharaja Ranbir Singh was a watershed moment in the region's water management history. This canal facilitated irrigation across vast agricultural lands, leading to increased agricultural productivity and economic growth.

In the modern era, Jammu witnessed significant developments in water management with the construction of dams and barrages. The Ranjit Sagar Dam, also known as Thein Dam, is a prominent example of such infrastructure. Completed in the early 2000s, it not only provides water for irrigation but also generates hydroelectric power, contributing to the region's energy needs.

However, despite these advancements, Jammu has faced challenges related to water management, such as the increasing demand due to population growth, inefficient irrigation practices, and environmental degradation. Climate change has also led to erratic rainfall patterns, impacting the region's water resources. Efforts to address these challenges have included the promotion of water conservation practices, the implementation of efficient irrigation techniques, and the construction of check dams to recharge groundwater.

In recent years, awareness about sustainable water management practices has grown, leading to community-led initiatives, government interventions, and the involvement of non-governmental organizations. These efforts aim to preserve traditional water sources, promote rainwater harvesting, and ensure the equitable distribution of water resources among different sectors.

The historical overview of water management in Jammu showcases the region's rich heritage of utilizing water resources for sustenance and growth. From ancient baoli's to modern dams, the evolution of water management techniques reflects the resilience and adaptability of the people of Jammu. As the region faces contemporary challenges, it is crucial to build upon the historical knowledge and continue implementing innovative and sustainable water management practices to secure a water-secure future for generations to come.

Addressing the challenges of water scarcity and contamination in Jammu requires a multi-faceted approach involving sustainable urban planning, efficient water management practices, strict regulation of industrial discharges, and community awareness initiatives. By taking proactive measures to conserve water, improve sanitation, and regulate pollutants, Jammu can mitigate the adverse effects of water scarcity and contamination, ensuring a healthier, more sustainable future for its residents.

Causes and Consequences

Water is essential for life, yet many regions, including Jammu, grapple with the twin challenges of water scarcity and contamination. Jammu, a city in the northern region of India, faces severe issues related to the availability and quality of water. This essay explores the causes and consequences of water scarcity and contamination in Jammu, shedding light on the pressing concerns that impact both the environment and the lives of its residents.

Main Causes of Water Contamination and Scarcity

1. Industrial pollution: Firstly Industrial pollution has emerged as a significant culprit behind the water contamination and scarcity issues in Jammu. The rapid industrialization in the region has led to the unchecked discharge of harmful chemicals, heavy metals, and toxins into water bodies. These pollutants seep into the groundwater, rendering it unsafe for consumption and agricultural use. Additionally, the excessive demand for water by industries has resulted in over-extraction from local water sources, depleting the natural reservoirs. The contamination not only jeopardizes the health of the residents but also disrupts the aquatic ecosystems, impacting the overall biodiversity. The unregulated industrial activities release harmful chemicals into water bodies, contaminating the primary source of our drinking water.

2. Agricultural Runoff: Secondly Agricultural runoff stands out as another major contributor to water contamination and scarcity issues in Jammu. The region's heavy reliance on agriculture has led to the widespread use of chemical fertilizers, pesticides, and herbicides. During rainfall, these chemicals are washed off fields and make their way into rivers and groundwater, contaminating water sources. The high levels of pollutants not only compromise the quality of drinking water but also harm aquatic life and disrupt the delicate balance of ecosystems. Furthermore, excessive irrigation practices often deplete water sources, leading to scarcity, especially during dry seasons.

3. Water disposal: Thirdly the Improper water disposal is also emerging as a significant cause of water contamination and scarcity in Jammu. Inadequate sewage systems and the haphazard disposal of waste, including untreated industrial effluents and domestic sewage, have severely polluted water bodies and groundwater. The untreated wastewater finds its way into rivers and lakes, contaminating them with harmful pathogens and pollutants. This contamination not only makes the water unfit for consumption but also poses serious health risks to the residents who rely on these sources. Additionally, the continuous discharge of untreated water into the environment exacerbates the scarcity issue by polluting existing water sources.

4. Deforestation: Fourthly deforestation in the region is also a critical factor contributing to water contamination and scarcity in Jammu. The rampant clearing of forests for agricultural expansion, urban development, and industrial activities disrupts the natural water cycle. Trees and plants play a vital role in stabilizing soil, preventing erosion, and promoting groundwater recharge. With the loss of forests, rainwater is not efficiently absorbed into the ground, leading to surface runoff. This runoff carries sediments, pollutants, and chemicals into water bodies, contaminating them and making the water unsafe for consumption and agricultural use. Moreover, deforestation reduces the overall availability of water by decreasing the capacity of the land to retain moisture, exacerbating the scarcity issue. Efforts to curb deforestation and promote afforestation are crucial to ensuring water security in Jammu, as they help maintain the balance of ecosystems and preserve the quality and quantity of water resources for the region.

5. Climate Change: Lastly Climate change has also been a driver of water contamination and scarcity in Jammu. Rising global temperatures and changing weather patterns have led to unpredictable rainfall, prolonged droughts, and altered precipitation cycles. These changes disrupt the natural flow and availability of water, leading to scarcity in certain seasons. Additionally, extreme weather events, such as heavy rainfall and floods, can overwhelm existing water infrastructure, contaminating water sources …

Consequences of Water Contamination and Scarcity

1. Public health crisis: First and foremost Water contamination and scarcity have severe consequences on public health, leading to a range of debilitating diseases and health challenges. Contaminated water sources, tainted with pollutants, bacteria, and chemicals, can cause waterborne illnesses such as cholera, dysentery, and gastroenteritis. These diseases result in severe dehydration, diarrhea, and, in extreme cases, can be fatal, especially among vulnerable populations such as children and the elderly. Additionally, water scarcity forces communities to rely on unsafe water sources, leading to a cycle of illness and poverty. Lack of access to clean water hampers basic hygiene practices, increasing the risk of infections and diseases. Moreover, limited water availability often leads to poor sanitation, inadequate wastewater management, and improper disposal of waste, further exacerbating health problems. Chronic water scarcity also impacts nutrition, as agriculture suffers due to lack of irrigation, affecting food supply and quality. Addressing water contamination and scarcity is vital not only for improving public health but also for promoting overall social and economic well-being in affected communities.

2. Agricultural impacts: Secondly Water contamination and scarcity have devastating consequences on agriculture, posing significant challenges to food production and livelihoods. Contaminated water, laden with chemicals and pollutants, harms crops by stunting growth, reducing yield, and compromising the quality of agricultural produce. Plants absorb these contaminants, leading to a decline in crop productivity and economic losses for farmers. Moreover, limited water availability due to scarcity hampers irrigation efforts, forcing farmers to rely on rain-fed agriculture, which is highly susceptible to climate variations. Insufficient water supply disrupts planting schedules and affects the overall agricultural calendar, leading to crop failures and reduced agricultural diversity. Additionally, water scarcity prompts farmers to use water-saving techniques such as drip irrigation, which, while helpful, may not be sufficient to sustain large-scale agriculture. This situation often perpetuates a cycle of poverty among farming communities, impacting their income and food security. Sustainable water management practices, coupled with investments in efficient irrigation systems, are essential to mitigate these consequences and ensure the resilience of agriculture in the face of water-related challenges.

3. Economic Challenges: Thirdly Water contamination and scarcity have far-reaching consequences on the economy of regions and nations. Contaminated water sources can lead to increased healthcare costs due to the treatment of waterborne diseases, putting a strain on public healthcare systems. Moreover, a workforce affected by illnesses caused by contaminated water is less productive, leading to a decrease in overall economic output. In agricultural economies, reduced crop yields due to water scarcity and contamination can result in income losses for farmers and drive up food prices in the market, impacting both producers and consumers. Industries reliant on water, such as manufacturing and tourism, may face higher operational costs or closures due to inadequate water supply, leading to job losses and reduced economic growth. Water scarcity can also spark conflicts over scarce resources, further destabilizing regions and hindering economic development. Additionally, investments in water infrastructure and clean-up efforts divert resources that could otherwise be used for education, research, and innovation. Addressing water contamination and scarcity is not just a matter of environmental concern but also a crucial step in promoting economic stability, public health, and sustainable development.

4. Environmental Degradation: Fourthly Water contamination and scarcity have bad consequences on the environment, disrupting delicate ecosystems and endangering biodiversity. Contaminated water sources harm aquatic life, leading to fish kills and the decline of species that depend on clean water for survival. Pollutants in water, such as heavy metals and chemicals, can accumulate in the food chain, posing risks to predators, including humans. Additionally, water scarcity puts stress on wetlands, lakes, and rivers, vital habitats for numerous plant and animal species. Reduced water levels can lead to habitat loss and migration challenges, threatening the existence of various aquatic and terrestrial organisms. Moreover, water scarcity exacerbates soil erosion and desertification, further degrading natural habitats.

On a larger scale, the imbalance in water availability and quality can disrupt the global climate system, as water plays a crucial role in regulating temperature and weather patterns. Addressing water contamination and scarcity is essential for preserving biodiversity, maintaining ecosystem services, and ensuring the long-term sustainability of our planet.

5. Social Strain: Last but not least Water contamination and scarcity have strained our social life, affecting communities in numerous ways. Contaminated water sources can lead to widespread illness and disease, straining healthcare systems and diminishing the overall quality of life. Families and communities spend significant resources on medical treatments, leaving them with less income for education and other essential needs. Additionally, the burden of fetching water often falls disproportionately on women and children, limiting their opportunities for education and personal development. Water scarcity can cause social tensions and conflicts, especially in regions where water resources are scarce. Communities might compete for limited water supplies, leading to disputes and strained relationships. Moreover, inadequate access to clean water and sanitation facilities can compromise personal hygiene, dignity, and overall well-being, affecting social interactions and community cohesion. Addressing water contamination and scarcity is crucial not only for safeguarding public health but also for fostering social equity, empowering communities, and promoting harmonious living conditions for all.

Proposed Solutions for Water Contamination and Scarcity in Jammu

The water crisis in Jammu demands urgent and strategic solutions to address both contamination and scarcity issues. This chapter explores comprehensive proposals aimed at ensuring clean and accessible water for the region's inhabitants. By implementing these solutions, Jammu can pave the way for a sustainable water future.

1. Advancing Water Treatment Plants: First and foremost Advancing water treatment plants stands out as a pivotal solution in the fight against water contamination and scarcity in Jammu. By investing in modern water treatment technologies and upgrading existing facilities, the region can significantly improve the quality of its water supply. Advanced treatment processes, including membrane filtration, chemical disinfection, and activated carbon absorption, are effective in removing contaminants such as bacteria, heavy metals, and chemicals. These technologies ensure that water meets stringent quality standards before reaching households. Moreover, the implementation of efficient water treatment methods enables the recycling and reuse of water, maximizing its utility and reducing wastage. Additionally, enhancing the capacity of treatment plants can address scarcity issues by optimizing water distribution, reducing leakages, and ensuring a more equitable supply to various communities. Public-private partnerships and community involvement can further support these initiatives, fostering sustainable water management practices. By advancing water treatment plants, Jammu can secure a clean, safe, and sustainable water supply, safeguarding public health, promoting agricultural growth, and fostering economic development in the region.

2. Fixing Sewage Systems: Secondly Addressing water contamination and scarcity in Jammu requires a comprehensive approach, with fixing sewage systems being a crucial proposed solution. Upgrading and expanding the existing sewage infrastructure is paramount to preventing untreated wastewater from entering water bodies. Properly functioning sewage systems not only prevent contamination of natural water sources but also ensure the safe disposal of human and industrial waste. By repairing and modernizing sewage systems, the region can significantly reduce the risk of waterborne diseases and improve overall public health. Furthermore, efficient sewage systems contribute to groundwater protection, preserving this vital resource for future use. Public awareness campaigns about responsible water usage and waste disposal, coupled with stringent enforcement of regulations, are essential to the success of this solution. By investing in sewage system repairs and maintenance, Jammu can mitigate water contamination, enhance the availability of clean water, and promote a healthier environment for its residents.

3. Advocating for Precision Farming: Thirdly Implementing precision farming techniques stands out as an innovative and effective solution in the fight against water contamination and scarcity in Jammu. Precision farming utilizes advanced technologies such as sensors, drones, and data analytics to optimize agricultural practices. By precisely monitoring soil moisture levels, crop health, and weather patterns, farmers can optimize irrigation, using water more efficiently and preventing overuse. This targeted approach not only conserves water but also minimizes the usage of fertilizers and pesticides, reducing the risk of chemical runoff contaminating nearby water sources. Precision farming promotes sustainable agriculture by ensuring that crops receive the right amount of water at the right time, leading to higher yields and decreased water wastage. Additionally, educating farmers about these modern techniques enhances their understanding of water conservation, contributing to a collective effort in mitigating water scarcity. By embracing precision farming, Jammu can revolutionize its agricultural practices, conserve precious water resources, and mitigate the risks of contamination, thereby promoting both environmental sustainability and food security for the region.

4. Encouraging Organic Farming: Fourthly Promoting organic farming emerges as a significant and eco-friendly solution in combating water contamination and scarcity in Jammu. Organic farming techniques emphasize natural methods, avoiding the use of chemical fertilizers, pesticides, and genetically modified organisms. By adopting organic practices, farmers can enhance soil fertility and structure, leading to improved water retention in the soil. Organic farming encourages the use of compost and organic matter, enriching the soil's ability to hold water, thus reducing the dependency on excessive irrigation. Additionally, avoiding chemical inputs mitigates the risk of groundwater contamination, ensuring cleaner water sources for both agriculture and local communities. Organic farming also nurtures biodiversity, including beneficial insects and microorganisms, which contribute to a healthier ecosystem. Furthermore, organic produce is in growing demand, offering economic incentives for farmers to transition away from conventional methods. By promoting organic farming, Jammu can conserve water resources, prevent contamination, and foster sustainable agricultural practices, ultimately contributing to a greener environment and a healthier populace.

5. Promoting Rainwater Harvesting: Fifthly Implementing rainwater harvesting techniques presents a practical and sustainable solution in addressing water contamination and scarcity in Jammu. By capturing rainwater, the region can reduce its dependency on traditional water sources, mitigating the strain on existing water supplies. Rainwater harvesting systems, including rooftop collection and storage tanks, enable communities to store rainwater during the monsoon season. This stored water can then be used for various purposes, such as irrigation, flushing toilets, and even drinking after proper filtration. By relying on rainwater, Jammu can decrease its demand on groundwater and surface water, preventing over-extraction and subsequent contamination. Moreover, rainwater is naturally pure, reducing the need for extensive chemical treatments. Implementing widespread rainwater harvesting not only conserves water but also provides a decentralized and reliable water source, empowering communities to manage their water needs effectively. Public awareness campaigns and incentives for implementing rainwater harvesting systems can further encourage its adoption, ensuring a more sustainable and resilient water supply for the region.

6. Roof-Top water Collection: Sixthly Promoting rooftop water collection stands as a practical and efficient solution in combating water contamination and scarcity in Jammu. By harnessing rainwater directly from rooftops, communities can significantly reduce their reliance on contaminated surface water and groundwater sources. Roof runoff, collected through gutters and directed into storage tanks, provides a clean and relatively pure source of water. This harvested rainwater can be treated and used for various non-potable purposes, such as watering gardens, flushing toilets, and cleaning. Implementing rooftop water collection not only conserves precious water resources but also helps prevent the contamination of existing water supplies. By encouraging households, businesses, and public institutions to adopt rooftop water collection systems, Jammu can create a decentralized water supply network, reducing the pressure on centralized sources and ensuring a more resilient and sustainable water future for the region. Public education campaigns and government incentives can play a vital role in encouraging widespread adoption of this eco-friendly practice.

7. Spreading Public awareness and education: Seventhly Elevating public awareness and education stands out as a pivotal solution in the battle against water contamination and scarcity in Jammu. By informing the community about the importance of water conservation, proper waste disposal, and responsible agricultural practices, individuals can make informed choices that reduce water pollution and promote sustainability. Public education campaigns can highlight the consequences of water contamination, emphasizing the link between clean water and overall public health. Moreover, educating farmers about efficient irrigation techniques and the judicious use of fertilizers and pesticides can significantly reduce agricultural runoff, preventing the contamination of water sources. Empowering the public with knowledge about rainwater harvesting, wastewater treatment, and the significance of preserving natural habitats can inspire collective action. Schools, community centers, and online platforms can serve as avenues for spreading this crucial information. Through a well-informed and engaged populace, Jammu can foster a culture of water conservation, ensuring the protection of its water resources for generations to come.

8. Putting Policies in Place: Eighthly Developing comprehensive policies and strengthening governance mechanisms are vital solutions in the fight against water contamination and scarcity in Jammu. Well-defined and effectively enforced policies can regulate industrial discharges, agricultural practices, and urban development, preventing harmful pollutants from entering water sources. Establishing stringent water quality standards and implementing regular monitoring can ensure that the water supplied to communities meets safety benchmarks. Additionally, investing in efficient sewage treatment plants, waste management systems, and upgrading aging infrastructure can significantly reduce the contamination of water bodies. Transparent and accountable governance is crucial for equitable water distribution, preventing illegal water extraction, and enforcing penalties for polluters. Collaborative efforts between government bodies, local communities, and environmental organizations can lead to the formulation of policies that promote sustainable water management. By prioritizing policy development and strengthening governance, Jammu can pave the way for a future where clean and accessible water is a reality for all, ensuring the well-being of its residents and the preservation of its natural resources.

9. Using Research and New Innovations: Lastly Harnessing new research and innovations offers a promising solution in the battle against water contamination and scarcity in Jammu. Investing in cutting-edge research can lead to the development of advanced water treatment technologies, efficient irrigation methods, and eco-friendly agricultural practices. Scientists and researchers can explore innovative solutions such as nanotechnology-based filtration systems, smart sensors for real-time water quality monitoring, and sustainable desalination techniques. Additionally, promoting research on drought-resistant crops and soil moisture management can aid farmers in optimizing agricultural productivity with minimal water usage. Embracing innovations like mobile apps for water conservation tips and digital platforms for monitoring water quality can enhance public awareness and engagement. Furthermore, encouraging startups and entrepreneurs to work on water-related innovations can lead to creative solutions that address specific local challenges. By fostering a culture of research and innovation, Jammu can stay ahead of the curve, adopting cutting-edge technologies to ensure a sustainable and secure water future for the region.

Researches & Results

This comprehensive research study delves into the pressing issue of water contamination and scarcity in Jammu, offering a detailed analysis based on the examination of sixty water samples sourced from various regions within the area. The research meticulously investigates the quality and availability of water resources, focusing on the impact of contamination on public health, agriculture, and the environment. Through rigorous testing and analysis, this study provides a thorough understanding of the challenges faced by Jammu concerning water quality and scarcity. The results, meticulously documented and presented in the attached reports, reveal critical findings that shed light on the extent of the problem. The research serves as a valuable resource, not only highlighting the severity of the issue but also offering essential data for policymakers, environmentalists, and local communities to devise effective strategies and sustainable solutions. By grounding the analysis in empirical evidence, this research aims to drive informed decision-making and prompt action towards ensuring a cleaner, safer, and more abundant water supply for the people of Jammu. Attached are all the reports with the findings.

CHANNI HIMMAT SEC -1

Parameter	Summer	Monsoon	Winter
Date	01/06/2022	15/07/22	16/11/22
Time	8:15 am	11:00 am	9:17 am
Location	Channi Himmat Supply 384-A/1		
Fecal Pollution(0)	Yes Fecal Contamination, E. Coli		

PERTH

Parameter	Summer	Monsoon	Winter
Date	30/06/2022	16/07/22	24/11/22
Time	3:12 pm	2:30 pm	3:00 pm
Collected by	Umer and Abhay	Dave	Abhishek
Location	Hand Pump at the edge of Road side village		
General Observations	Large pump with uneven, slight flow; lager permanent settlement		
Litter	Large amount on road side just outside settlement		
Feces	Buffalo		
Household effluents	Soap or detergent noted in pool and TW1 off form source		
Factory / industry effluents	Road paving both up and down road of settlement		

Parameter	Summer	Monsoon	Winter
Animals	Large buffalo tied up at source		
Washing directly at the water	Soap/detergent noted see above		
Water usage	Runoff from source ran into Heavily littered road side ditch. No usage noted.		
pH(6.5-8.5)	9	8.5	9
Dissolved Oxygen(5)	6.2	6.2	6.0
Turbidity (5-10)	10	8	9
Nitrate (45max)	45	10<N<45	<10
Phosphorous	<.I		
Fecal pollution(0)	No	No	No
Residual	<.2	<0.2	<0.2

CHANNI HIMMAT

Parameter	Summer	Monsoon	Winter
Date	01/06/22	02/08/22	03/10/22
Time	11:00 am	12:35 pm	9:26 am
Collected by	Perry	Abhishek	Sohan
Location	Channi Himmat Sector 5 empty lot		
General Observations	Water taken from a leak/crack in the municipal water supply ptpe. Compliant of varying water quality, Sometimes seems like 'Soap'		
pH(6.5-8.5)	7	8	7
Dissolved Oxygen(5)	1.6mg/I	1.7 mg/l	1.6 mg/l
Nitrate(45max)	<10mg/l	<10mg/l	<10mg/l
Phosphorous	0.5mg/I	0.5mg/I	1.7mg/I
Fecal pollution(0)	Yes	Yes	Yes
Residual Chlorine(0.2)	<0.2mg/l	0	0
Chloride (250-1000)	70.9mg/I	35.4	70.9mg/I

Parameter	Summer	Monsoon	Winter
Fluoride (1-1.5)	0.6mg/I	<0.6	0.6mg/I
Iron(0.3)	0.2	0.3	0.4
Ammonia (1.5)	<1.0mg/I	<1.0	<1.0
Arsenic (less than10 ppb)	Positive 50ppb	Positive 10ppb	Positive 10ppb

SANDARANI I

Parameter	Summer	Monsoon	Winter
Date	30/05/2022	24/07/22	03/02/22
Time	11:30 am	01:09 pm	12:55 pm
Collected by	Janet/Cini	Abu, Charan	Abhishek
Location	Pipe runs from spring above house		
Environmental Conditions	Normal 99 degrees		
General Observations	Agriculture uphill, this house was down hill		
Water Usage Description :	Family said that pipe was their source for every thing		
pH(6.5-8.5)	7.5 units	8.5	8
Dissolved Oxygen (5)	14mg/I	14mg/I	14mg/I
Turbidity(5-10)	7	8	4
Nitrate (45max)	10mg/1	10	10
Phosphorous	0mg/1		NVR
Fecal Pollution (0)	Yes Fecal Contamination, E.Coli	Yes Fecal Contamination, E.Coli	No

Parameter	Summer	Monsoon	Winter
Residual Chlorine (0.2)	<0.2mg/l	0	NVR
Chloride (250-1000)	63.81mg/I	35.4	106.35
Fluoride(1-1.5)	0.6mg/I	0.6	1
Iron(0.3)		<0.3	
Ammonia(1.5)	<1.0mg/I	<1	1
Arsenic(I0ppb)	Positive 200 ppb	Absent	Absent
GPS Coordinates	N3258'27.73E7459'25.69	Same	Same

SANDARANI II

Parameter	Summer	Monsoon	Winter
Date	30/05/2022	24/07/22	03/01/22
Time	11:50 am	3:44 pm	5:24 pm
Code No.	2		
Collected by	Janet and Cini	Charan	Abhishek
Location	Family gets water at the temple		
Environment Conditions	Open spring		
Litter	Yes	Yes	Yes
Feces	Yes		
Animals	Yes, drinking		
Riverbank Vegetation	Yes		
Agricultural runoff	Yes		
Washing directly at the water	Yes, 2 men bathing at time of Sampling		
Water] usage description:	Bathing, drinking, washing		
pH(6.5-8.5)	8	9	9
Dissolved Oxygen (5)	9.6	9.7	9.6

Parameter	Summer	Monsoon	Winter
Turbidity(5-10)	I0 mg/l	6 mg/l	8 mg/l
Nitrate(45max)	<10	<10	<11
Phosphorous	<0.1		<.I
Hardness (3000-600)	600	700	600
Fecal Pollution (0)	Yes, Fecal Contamination, E.Coli	Yes	Yes
Residual chlorine	<0.2	0	NVR

SANDARANI III

Parameter	Summer	Monsoon	Winter
Date	30/05/2022	24/07/22	30/02/22
Time	12:00 pm	01:55 pm	10:11 am
Collected by	Janet and Cini	Sohan, Charan	Abhishek
Location	House with blue water tank, Hauled from far away		
Feces	Yes		
Water usage description	Bathing, drinking, washing		
pH(6.5-8.5)	8.5	8.5	8
Dissolved Oxygen(5)	8	8	8
Turbidity(5-10)	<10	<10	<10
Nitrate(45max)	<10	<10	<10
Phosphorous	<0.1	<0.1	<0.1
Fecal Pollution(0)	Yes, Fecal Contamination, E. coli	Yes	No
Residual Chlorine(0.2)	<0.2	0	NVR
Chloride (250-1000)	46.085	35.4	106.35

Parameter	Summer	Monsoon	Winter
Fluoride(1-1.5)	0.6	0.6	1
Fluoride(1-1.5)	0.6	0.6	1
Iron(0.3)	0.3	0.3	0.3
Ammonia(1.5)	<1.0	1	I
Arsenic(10ppb)	Positive 50 ppb	Positive 50 ppb	Absent

BALI NALA

Parameter	Summer	Monsoon	Winter
Date	01/06/2022	24/08/22	24/02/22
Time	8:25 am	10:26 am	3:50 pm
Collected by	Matt	Umar, Charan	Abhishek
Location	Dried up water fill, pools remain		
Environmental conditions	Rocky, hot		
General Observations	Rocky area was a waterfall but dried up		
Litter	Yes		
Bank Stability/erosion	Yes		
Water usage description	Drinking		
pH(6.5-8.5)	8.5units	8.5	7.5
Dissolved Oxygen(5)	I0mg/1	I0mg/1	I0mg/1
Turbidity(5-10)	< 10	< l0	< 10
Nitrate(45max)	10mg/1	<10	10
Phosphorous	<.01mg/I	<0.1mg/I	<0.2 mg/I

Parameter	Summer	Monsoon	Winter
Fecal pollution(0)	Yes	Yes	Yes
Residual chlorine(0.2)	<.2mg/I	0	NVR
Chloride (250-1000)	283.6mg/I	35.4	106.35
Fluoride(1-1.5)	0.6mg/I	0.6	1
Iron(0.3)	0.3	0.3	0.3
Ammonia(1.5)	1.0mg/1	<1.0	<I
Arsenic(l0ppb)	100ppb	100ppb	Absent
GPS Coordinates	N32,58.524',E75 10.527'	Same	Same

BALI NALA MASJID

Parameter	Summer	Monsoon	Winter
Date	01/06/2022	25/07/22	03/01/22
Time	12:40 pm	9:00 am	2:35 pm
Collected by	Sohan	Abhishek	Umar
Location	From water tank (different source)		
General Observations	Near pad by shrine/temple, come from runoff from above then collected 10 tank.		
Litter	Yes		
Animals	Yes		
Bank stability/erosion	Yes		
Water usage description:	Drinking, Washing		
pH(6.5-8.5)	9	8.5	7
Dissolved Oxygen(5)	8	5	8
Turbidity(5-10)	5	6	9
Nitrate(45max)	<10	<10	NVR
Phosphorous	<0.1	<0.1	<0.1

Parameter	Summer	Monsoon	Winter
Fecal pollution(0)	Yes, Fecal Contamination, E. Coli	Yes	No
Residual chloride(0.2)	<0.2	0	NVR
Chloride (250-1000)	70.9	42.5	106.35
Fluoride(1-1.5)	0.6	0.6	1
Iron(0.3)	0.3	0.3	0.3
Ammonia(1.5)	1	<1	<1
Arsenic(l0ppb)	l00ppb	50ppb	Absent

Water Truck City Supply

Parameter	Summer	Monsoon	Winter
Date	31/05/2022	20/06/22	19/11/22
Time	3:00 pm	9:57 am	1:19 pm
Code No.	Truck water pumped from Narwal or Sainik Colony		
Fecal pollution(o)	Yes, Fecal Contamination, E. Coli	Yes	No

JAMODA

Parameter	Summer	Monsoon	Winter
Date	17/05/2022	18/08/22	26/01/22
Time	2:35 pm	1:00 pm	11:00 am
Collected by	Binnu and Marko	Umar	Abhishek
Location	Contact claimed torely on rainwater		
Environmental conditions	Mountain top, sparse vegetation, hot and hazy		
General Observations	Small group of less than 10 houses On top of a rocky mountain side		
Litter	Yes		
Feces	Yes animal and human		
Water usage description:	No visual bucket or tanks to capture rain water		
pH(6.5.8.5)	7	8.5	7
Dissolved Oxygen(5)	80	70	50
Turbidity(5-10)	<10	<11	No
Nitrate(45max)	<10	<10	NVR
Phosphorous	0.35	0.35	0.40
Fecal pollution(0)	Yes, Fecal Contamination, E. Coli	Yes	Yes

Parameter	Summer	Monsoon	Winter
Residual Chlorine(0.2)	<0.7	<0.2	<0.3
Chloride (250-1000)	70.9	53	250
Fluoride(1-1.5)	< 0.1	< 0.6	< 1.5
Iron(0.3)	0.2	3	0.3
Ammonia(1.5)	< 3	<1.0	<0.3
Arsenic(I0ppb)	Negative	Negative	Negative

BUPNER GARH

Parameter	Summer	Monsoon	Winter
Date	27/05/2022	29/08/22	30/12/22
Time	1:43 pm	4:35 pm	10:55 pm
Collected by	Marko	Abhishek	Umar / Sohan
Location	Tank pumped from lake; High elevation in state park		
General Observations	Many tourists around lake; Fish for turtles		
Litter	In lake		
Faeces	Yes		
Factory/industry influents	Reported chlorine treatments		
Water usage description:	Bathing, drinking, washing, irrigation, recreation		
pH(6.5-8.5)	8	7.5	8
Dissolved Oxygen(5)	16.8mg/l	12.4mg/l	16.8mg/l
Turbidity(5-10)	<l0NTIJ	<l0NTIJ	<l0NTIJ
Nitrate(45max)	<10mg/1	<10	<10
Phosphorous	<0.1mg/1	Same	Same

Parameter	Summer	Monsoon	Winter
Faecal pollution(0)	Yes, Fecal Contamination, E. Coli	Yes	Yes
Residual Chlorine(0.2)	<0.2mg/l	<0.2	<0.2
Chloride (250-1000)	53.175mg/l	63	250
Fluoride(1-1.5)	1mg/l	6 mg/l	1mg/l
Iron(0.3)	0.3	0.3	0.4
Ammonia(1.5)	1.0mg/l	1.0 mg/l	1.0mg/l
Arsenic(10ppb)	Negative	Negative	Negative

INDRA NAGAR

Parameter	Summer	Monsoon	Winter
Date	09/06/2022	12/07/22	26/02/22
Time	10:40 am	2:00 pm	6:00 pm
Collected by	Marco, Lindsey, Joy	Perry	Abhishek
Location	Hand pump in dera		
Environmental conditions	Hot, clear		
Litter	Some present		
Feces	Large piles Animal Dung		
Animals	Large herd of buffalo in river		
Habitat destruction	Road construction		
pH(6.5-8.5)	8	8.5	8
Dissolved Oxygen(5)	4	4.5	3.5
Turbidity	<10	<10	<10
Nitrate(45max)	10	<10	45
Phosphorous	0.1	0.7	NVR
Fecal pollution(0)	No	Yes	No

Parameter	Summer	Monsoon	Winter
Residual chlorine(0.2)	<0.2	<0.2	NVR
Chloride (250-1000)	212.7	88	212.7
Fluoride(1-1.5)	0.6	<0.6	0.6
Iron(0.3)		1	
Ammonia(1.5)	<1	1	NVR
Arsenic(10ppb)	Positive 100ppb	Negative	Negative

DAK BANGLA MASJID

Parameter	Summer	Monsoon	Winter
Date	13/06/2022	08/07/22	16/11/22
Time	10:55 am	5:00 pm	10:26 am
Collected by	BPJ/MSA	Perry	Abhishek
Location	Dak Bangla electric pump		
General observations	Electric pump, walled in		
Feces	Dung and fecal patties		
Animals	Large herd of buffalo and goats		
Washing directly at the water	Man bathing		
Water usage description:	Large Stagnant, green pool with large dung pile right next to it and some in it. No Activities noted in the pool.		
pH(6.5-8.5)	7.5	7.5	7
Nitrate(45max)	100	45<N<100	45
Phosphorous	0.5		NVR

Parameter	Summer	Monsoon	Winter
Fecal pollution(0)	No	Yes	No
Residual chlorine(0.2)	<0.2	<0.2	NVR
Chloride (250-1000)	141.8	53	106.35
Fluoride(l-1.5)	1.2	<0.6	0.6
Iron (0.3)	1	1	0.3
Ammonia(1.5)	1	1.5	1
Arsenic(10ppb)	Positive 50ppb	Absent	50ppb

RAKH BARUTI

Parameter	Summer	Monsoon	Winter
Date	30/05/2022	23/07/22	25/2/22
Time	3:35 pm	04:00 pm	1:00 pm
Collected by	Marco/ Lindsay	Umar, Charan	Abhishek, Sohan
Location	The near side of vijaypur Gujjar Basti Salam Din		
General observations	Buffalo herd; mud and staw		
Litter	Out skirt of village		
Feces	Buffalo		
Animals	Buffalo at trough		
Washing directly at the water	At pump; bathing and clothing		
Water usage description	Stagnant pond; drains Down		
pH(6.5-8.5)	7units	8.5	7
Dissolved oxygen(5)	2mg/I	2mg/I	4mg/I
Turbidity	40NTU	40	40
Nitrate(45max)	45mg/I	10	10

Parameter	Summer	Monsoon	Winter
Phosphorous	0.1mg/l	0.1mg/l	0.1mg/l
Faecal pollution(0)	No	No	No
Residual chlorine(0.2)	0.0mg/l	0.2	NVR
Chloride(250-1000)	70.9mg/l	35.4	106.35
Fluoride(1-1.5)	0.6mg/I	<0.6	0.6
Iron (0.3)	0.3	1	1
Ammonia(1.5)	1.0mg/1	1	1
Arsenic	Negative	Negative	Negative

BADIYA

Parameter	Summer	Monsoon	Winter
Date	30/06/2022	13/07/22	15/02/22
Time	9:26 am	11:12 am	5:39 pm
Collected by	Binnu Marko	Abhishek, Sohan	Umar, Abhishek
Location	Bah-Di-Uh The far side (toward Samba) of Vijaypur Gujjar Basti Hand pump Middle of small village		
Feces	Animal feces around tank		
Factory/Industry effluents?	Contact claimed nearby distillery polluted water. Could skim pollutants off top of water.		
pH(6.5-8.5)	8	8.5	8
Dissolved Oxygen(5)	11	4	4.5
Turbidity(5-10)	25	25	25
Nitrate(45max)	<10	<10	<10
Phosphorous	0.1	0.1	1
Faecal pollution(0)	Yes, Fecal Contamination, E. Coli	No	No

Parameter	Summer	Monsoon	Winter
Residual chlorine(0.2)	0.1	0	<0.2
Chloride(250-1000)	177.2	42.5	177.25
Fluoride(l-1.5)	1	<0.6	<0.6
Iron(0.3)	0.3	1	1
Ammonia	I	<1.0	<1.0
Arsenic(10ppb)	Positive 200ppb	Absent	Absent

BATALI

Parameter	Summer	Monsoon	Winter
Date	02/06/2022	22/07/22	25/02/22
Time	2:05 pm	1:00 pm	3:00 pm
Collected by	Andrea, Lindsay, Sonya	Abhishek, Caitlin	Charan & Saleem
Location	Hand pump		
Environmental conditions	Hot, dry		
General observations	Factory smoke nearby, Building under construction		
pH(6.5-8.5)	7.5	9	8.5
Dissolved oxygen(5)	8	5	6
Turbidity(5-10)	Less than 10	Same	Same
Nitrate(45max)	Less than 10	<10	NVR
Phosphorous	0.1	0.1	NVR
Faecal pollution(0)	No	No	No
Residual chlorine(0.2)	Less than 0.2	0	<.2
Chloride (250-1000)	177.25	35.4	70.9

Parameter	Summer	Monsoon	Winter
Fluoride(I-1.5)	0.6	20.6	<.6
Iron(0.3)	0.3	0.3	0.3
Ammonia(l.5)	1	3	<.1
Arsenic(10ppb)	Absent	Absent	Absent

TAHRA

Parameter	Summer	Monsoon	Winter
Date	30/05/2022	29/07/22	25/01/22
Time	4:39 pm	5:26 pm	10:04 am
Collected by	Dave	Abhishek, Charan	Sohan
Location	Natural spring out in Open		
Environmental conditions	Hot, dry		
General observations	Looks like a cave with Water, under a tree		
Litter	Yes		
Feces	Yes		
Animals	Fish in water source		
Bank stability/erosion	Yes		
Water usage description:	Drinking is all we know		
pH(6.5-8.5)	8unites	8.5	7.5
Dissolved oxygen(5)	6mg/l	4mg/l	4.5mg/l
Nitrate(45max)	<10mg/1	<10mg/1	NVR

Parameter	Summer	Monsoon	Winter
Phosphorous	0.1mg/l	0.2 mg/l	0.1mg/l
Faecal pollution(0)	Yes, Fecal Contamination, E. Coli	Yes	Yes
Residual chlorine(0.2)	<0.2mg/l	<0.2mg/l	<0.4mg/l
Chloride (250-1000)	70.9mg/l	42.5	200
Fluoride(1-1.5)	0.6mg/l	0.6mg/l	0.6mg/l
Iron(0.3)	<0.3	<0.3	<0.3
Ammonia(1.5)	<1.0mg/l	<1.0mg/l	<1.0mg/l
Arsenic	Absent	Absent	Absent

SURTA

Parameter	Summer	Monsoon	Winter
Date	01/06/2022	29/07/22	27/2/2022
Time	10:00 am	11:00 pm	1:26 pm
Collected by	Abu	Abhishek, Charan	Matt, Abhishek
Location	Temple		
General observations	Down by river, cement Tank and pipe from above		
Litter	Yes		
Feces	Yes		
Animals	Yes		
Bank stability/ erosion	Yes		
Washing directly at the water?	Yes		
Water usage description:	Bathing drinking washing		
pH(6.5-8.5)	9	9	7.5
Dissolved oxygen(5)	8	5	45

Parameter	Summer	Monsoon	Winter
Nitrate (45max)	15	10	10
Phosphorous	0.1	0.1	NVR
Faecal pollution(0)	Yes	Yes	No
Residual chlorine	<0.2	<0.2	NVR
Chloride (250-1000)	88.6	42.48	106.35
Fluoride(1-1.5)	0.6	0.6	About1.0
Iron(0.3)	<0.3	<0.3	<0.3
Ammonia(1.5)	1	<1	1.5
Arsenic (10ppb)	Absent	Absent	Absent

SOLARA DUBARR

Parameter	Summer	Monsoon	Winter
Date	30/05/2022	29/07/22	03/01/22
Time	10:00 am	1:00 pm	3:00 pm
Collected by	Janet and Cini	Abhishek	Umar
Location	White rock river bed. Spring ½Mile		
Litter	Yes		
Feces	Yes		
Household Effluents			
Water usage Description:	Bathing, washing, cooking, drinking		
pH(6.5-8.5)	7	9	7
Dissolved Oxygen (5)	12	4	5
Turbidity (5-10)	<10	<5	<5
Nitrate (45max)	<45	<10	<10
Phosphorus	0.1	0.1	0.5

Parameter	Summer	Monsoon	Winter
Faecal pollution(0)	Yes, Fecal Contamination E. Coli	Yes	No
Residual chlorine(0.2)	<0.2	<0.2	NVR
Chloride (250-1000)	111.25	56.6	106.35
Fluoride (1-1.5)	0.6	0.6	1
Iron(0.3)	<0.3	<0.3	<0.3
Ammonia(1.5)	3	1	1
Arsenic (10ppb)	Negative	Negative	Negative

TIHATI

Parameter	Summer	Monsoon	Winter
Date	30/05/2022	14/08/22	20/03/22
Time	1:35 am	2:00 pm	3:00 pm
Collected by	Janet and Cini		
Location	Drove through river to get to house. Not sure where Source is because boys were sent to get sample.		
Litter	Dead chick in house. Dirty floors with ants		
Water usage Description:	River just 50ft away. Everyone in river washing cars, bodies, playing etc.		
pH(6.5-8.5)	9	7	7
Dissolved Oxygen(5)	10	4	5
Turbidity (5-10)	10 mg/l	6 mg/l	10 mg/l
Nitrate (45max)	<10	<10	<10

Parameter	Summer	Monsoon	Winter
Phosphorous	<0.1	<0.1	<0.1
Faecal pollution(0)	Yes, fecal contamination, E. Coli	No	No
Residual chlorine(0.2)	<0.2	<0.2	<0.2
Chloride (250-1000)	111.25	56.6	106.35
Fluoride (1-1.5)	0.6	0.6	0.6
Ammonia(1.5)	<1	<1	<1
Arsenic	Positive 20ppb	Negative	Negative

JHANAKHA I

Parameter	Summer	Monsoon	Winter
Date	30/05/2022	26/07/22	27/02/22
Time	10:00am	12:05 pm	1:00 pm
Collected by	Matt	Abhishek	Matt, Abhishek
Location	Cement spring by natural water		
Environmental conditions	Hot, dry		
General observations	Deep cement hole next to natural Water hole....woman had fever, man had back aches		
Litter	Open, exposed		
Feces	Yes		
Animals	Yes		
Bank stability/ Erosion	Yes		
Agricultural runoff	Fanning near		
Washing directly at the water	Drinking, bathing, animals		

Parameter	Summer	Monsoon	Winter
pH(6.5-8.5)	8 units	8.5	8
Turbidity(5-10)	I0NTU	Same	Same
Nitrate(45max)	<10mg/l	<10mg/l	NVR
Phosphorus	0.1mg/l	Same	Same
Faecal pollution(0)	Yes	Yes	Yes
Residual chlorine(0.2)	<0.2mg/1	<0.2	NVR
Chloride (250-1000)	106.35mg/I	38.9	106.35
Fluoride (1-1.5)	0.4mg/I	<0.6	0.6
Iron(0.3)	0.3	0.3	0.3
Ammonia(1.5)	0.8mg/1	1	1
Arsenic	Negative	Negative	Negative

JHANAKHA II

Parameter	Summer	Monsoon	Winter
Date	30/05/2022	29/07/22	27/02/22
Time	9:00 am	12:05 pm	4: 01 pm
Collected by	Janet, Cini	Abhishek	Sohan, Umar
Location	½ Km to water source a Spring uphill		
General Observations	Used by old man - sent kid to go get water		
Feces	Yes		
pH(6.5-8.5)	9	9	7.5
Dissolved Oxygen(5)	80	80	70
Turbidity(5-10)	<10	9	7
Nitrate(45max)	<10	<10	NVR
Phosphorous	0.1	0.1	NVR
Faecal pollution(0)	Yes, Fecal Contamination, E. Coli	Yes	Yes
Residual Chlorine (0.2)	<0.2	<0.2	NVR
Chloride (250-1000)	177.25	49.5	106.35

Parameter	Summer	Monsoon	Winter
Fluoride (1-1.5)	0.6	0.6	0.6
Iron(0.3)	0.2	0.3	0.3
Ammonia(1.5)	1	1	1
Arsenic(I0ppb)	10ppb	10ppb	10ppb

CHAK

Parameter	Summer	Monsoon	Winter
Date	30/05/2022	29/07/22	14/12/22
Time	12:15pm	1:00pm	3:00pm
Collected by	Janet, Cini	Sohan	Abhishek
Location	½Km to water source a spring uphill		
Environmental conditions	Sunny		
Animals	Yes		
Riverbank vegetation	Grass		
Agricultural runoff	Yes, beside farm area		
pH(6.5-8.5)	7	8.5	8.5
Dissolved Oxygen(5)	10	Same	Same
Nitrate(45max)	30	10	NVR
Phosphorus	15	14	15
Faecal Pollution(0)	Yes, Fecal Contamination, E. Coli	No	Yes
Residual Chloride(0.2)	Less than 0.2	Same	Same
Chloride(250-1000)	70.9	42.5	200

Parameter	Summer	Monsoon	Winter
Fluoride(1-1.5)	0.6	0.6	0.6
Iron(0.3)	0.3	0.3	0.3
Ammonia(1.5)	1	1	1
Arsenic (I0ppb)	Negative	Negative	Negative

NAGROTA BYPASS

Parameter	Summer	Monsoon	Winter
Date	30/05/2022	29/07/22	16/11/22
Time	1:10 pm	2:20 pm	3:00pm
Collected by	Janet, Cini	Umar, Charan	Abhishek, Sohan
Location	½ Km to water Source a spring uphill		
Environmental conditions	Hot weather dry		
General observations	Along bypass; Not Easily visible		
Feces	No		
Bank Stability /erosion	Yes		
Washing directly at the water	Yes		
Water usage description:	Bathing,washing, drinking		
pH (6.5-8.5)	8	7	7
Nitrate (45max)	0	0	NVR

Parameter	Summer	Monsoon	Winter
Phosphorus	0	0	0
Faecal pollution(0)	Yes	Yes	Yes
Residual pollution	0	0	0
Chloride (250-1000)	110.5	35.4	60.5
Fluoride(1-1.5)	0.5mg/I	0.5mg/I	0.7mg/I
Iron(0.3)	0.3	0.3	0.2
Ammonia(1.5)	0	1.5	1.5
Arsenic(1Oppb)	Negative	Negative	Negative

MAZINE GANGYAL

Parameter	Summer	Monsoon	Winter
Date	30/5/2022	17/07/22	26/2/22
Time	12:35 pm	4:00 pm	3:00 pm
Collected by	Dave and Matt	Perry	Abhishek Charan
Location	Out of Tank		
Environmental conditions	Rocky, dry, hot		
General observations	Village right off the Road and down the hill		
Feces	Animals around house		
Animals	Yes		
Water usage description:	Drinking, washing, Bathing		
pH(6.5-8.5)	7	7.5	7
Dissolved Oxygen(5)	8mg/l	8mg/l	8mg/l
Turbidity(5-10)	4	7	8
Nitrate(45max)	45	40	42
Phosphorus	0.1mg/l	0.1mg/l	0.1mg/l
Faecal pollution	Yes	Yes	no

Parameter	Summer	Monsoon	Winter
Residual Chlorine(0.2)	<0.2mg/1	<0.2	<.2
Chloride(250-1000)	141.8mg/1	70	106.35
Fluoride(1-1.5)	<0.6	<0.6	<0.6
Iron(0.3)	0.3	0.3	0.3
Ammonia(1.5)	1.0mg/1	1	NVR
Arsenic	Positive 200ppb	Negative	Negative

RAVIA NAWAL

Parameter	Summer	Monsoon	Winter
Date	02/06/2022	15/07/22	04/01/22
Time	1:10 pm	2:00 pm	3:00 pm
Collected by	Sonya, Lindsay, Andree	Abhishek	Abhishek
Location	Road side faucet		
Environmental conditions	Hot, Dry		
General Observations	Gov't source, on all day		
pH(6.5-8.5)	8	7	6
Dissolved Oxygen(5)	10	10	5
Nitrate(45 max)	10	40	40
Phosphorus	0.1	0.1	0.1
Faecal pollution(0)	Yes, Fecal Contamination, E. Coli	Yes	No
Residual Chlorine(0.2)	Less than 0.2	Less than 0.2	Less than 0.2
Chloride (250-1000)	70.9	110.5	605.1
Ammonia(1.5)	Less than 1	Less than 1	Less than 1
Arsenic(10ppb)	Negative	Negative	Negative

KATLAYE SAMBA

Parameter	Summer	Monsoon	Winter
Date	02/06/2022	02/07/22	27/02/22
Time	12:30 pm	1:00 pm	2:00 pm
Collected by	Andrea, Lindsay, Soniya	Abhishek	Charan & Saleem
Location	Handpump		
Environmental conditions	Hot, dry		
General observations	Clean, small, garden		
pH(6.5-8.5)	8	8.5	8
Dissolved Oxygen(5)	8	8	8
Nitrate(45max)	Less than 10	10	<10
Phosphorus	0.3		0.1
Faecal pollution	No	No	No
Residual chlorine(0.2)	Less than 0.2	0	NVR
Chloride(250-1000)	70.9	35.4	106.35
Fluoride(1-1.5)	0.6	<0.6	0.6
Iron(0.3)	0.3	1	0.3

Parameter	Summer	Monsoon	Winter
Ammonia(1.5)	1	2	1
Arsenic	Negative	Negative	Negative

RAKHAM BATALI

Parameter	Summer	Monsoon	Winter
Date	30/05/2022	16/07/22	21/02/22
Collected by	Binnu	Abhishek	Abhishek
Location	Well near a cluster of houses		
General Observations	Small of cluster of permanent and grass huts, small hamlet, less than 12 people present, clean, Hindu And Muslim families		
Feces	Animal		
Animals	Cows dogs chickens		
Water usage description:	Stagnant pool near the well		
pH(6.5-8.5)	9	8.5	9
Dissolved oxygen(5)	10.2	10.2	10.2
Turbidity(5-10)	<10	<10	<10
Nitrate(45max)	<10	<10	NVR
Phosphorus	0.5	0.5	0.5
Faecal Pollution(0)	No	No	No
Residual Chlorine(0.2)	<0.2	0	0

Parameter	Summer	Monsoon	Winter
Chloride (250-1000)	779.9	35.4	300.6
Fluoride(1-1.5)	1	<0.6	1
Iron(0.3)	1	0.3	1
Ammonia(1.5)	1	3	1
Arsenic	Negative	Same	Same

BADGOO (BADSOOL)

Parameter	Summer	Monsoon	Winter
Date	02/01/2022	27/07/22	28/01/22
Time	10:00 am	11: 00 am	12:01 pm
Collected by	Dave	Abhishek	Charan
Location	Well pump		
pH(6.5-8.5)	8	9	9
Dissolved Oxygen(5)	6	6	6
Nitrate(45Max)	10	<10	NVR
Phosphorous	Less than 0.1	0	0
Faecal Pollution (0)	No	Yes	Yes
Residual Chlorine(0.2)	Less than 0.2	<0.2	10.2
Chloride(250-1000)	35.45	63.7	50.7
Fluoride(1-1.5)	0.6	0.6	0.7
Iron(0.3)	0.3	0.3	0.3
Ammonia(1.5)	Less than 1	<0.1	<0.1
Arsenic(10ppb)	10ppb	Negative	Negative

DHANSAL

Parameter	Summer	Monsoon	Winter
Date	02/01/2022	27/07/22	27/03/22
Time	10:00 am	11:00 am	12:00 pm
Collected by	Dave, Marco	Abhishek	Umar, Charan
Location	A well behind a Shop		
Environmental conditions	Sunny		
General observations	Shaded well about 100/20190 feet deep.		
pH(6.5-8.5)	8	9	8
Dissolved Oxygen(5)	6	6	5
Turbidity(5-10)	<10	<6	<8
Nitrate(45Max)	10	10	NVR
Phosphorous	0.1	0.1	0.1
Faecal Pollution(0)	No	No	No
Residual Chlorine(0.2)	<0.2	<0.2	<0.1

Parameter	Summer	Monsoon	Winter
Chloride (250-1000)	106.35	77.8	79.2
Fluoride(1-1.5)	0.6	0.5	0.5
Iron(0.3)	0.3	0.3	0.3
Ammonia(1.5)	<1	<1	<1
Arsenic(10ppb)	Negative	Negative	Negative

JHAJHAR KOTLI

Parameter	Summer	Monsoon	Winter
Date	06/01/2022	27/07/22	27/02/22
Time	10:00 am	11:00 am	1:00 pm
Collected by	Abu and Matt	Sohan	Sohan
Location	Handpump near road		
General Observations	Right off road by Shop		
Litter	Yes		
Feces	No		
Household effluents	No		
Factory *I* Industry Effluents	No		
Animals	Yes		
Agricultural runoff	Farm area near by		
pH(6.5-8.5)	8	9	7.5
Dissolved Oxygen(5)	8	5	8
Nitrate(45Max)	<10	<10	NVR

Parameter	Summer	Monsoon	Winter
Phosphorous	<0.1	<0.1	NVR
Faecal Pollution(0)	No	No	No
Residual Chlorine(0.2)	<0.2	<0.2	NVR
Chloride(250-1000)	70.9	35.4	106.35
Fluoride(250-1000)	0.6	0.6	0.6
Iron(0.3)	0.3	0.3	0.3
Ammonia(1.5)	1	<1	<1
Arsenic(10ppb)	10ppb	Negative	Negative

DUMEL SLERTH

Parameter	Summer	Monsoon	Winter
Date	06/01/2022	27/07/22	15/03/22
Time	10:00 am	12:01 pm	3:00 pm
Collected by	Dave, Matt, Andrea	Sohan, Charan	Abhishek, Dave
pH(6.5-8.5)	8.5	8.5	8.5
Dissolved Oxygen(5)	4	4	4
Turbidity(5-10)	<10	<10	<10
Nitrate(45Max)	<10	<10	<10
Phosphorous	<0.1	<0.1	<0.1
Faecal Pollution(0)	No	Yes, Fecal Contaminatio, E. Coli	Yes
Residual Chlorine(0.2)	<0.2	<0.2	<0.2
Chloride(250-1000)	177.25	35.4	177.30
Fluoride(1-1.5)	0.6	0.6	0.6
Iron(0.3)	0.3	0.3	0.3
Ammonia(1.5)	1.5	1.5	1.5
Arsenic	Negative	Negative	Negative

KANYALA

Parameter	Summer	Monsoon	Winter
Date	06/01/2022	31/07/22	26/03/22
Time	10:00 am	12:05pm	1:00 pm
Collected by	Dave	Sohan, Charan	Abhishek
Location	Handpump		
pH(6.5-8.5)	7	9	9
Dissolved Oxygen(5)	7	5	7
Nitrate(45Max)	0	<10	NVR
Phosphorous	1	1	1
Faecal Pollution(0)	No	Yes	Yes
Residual Chlorine(0.2)	<02	<0.2	<0.2
Chloride(250-1000)	212.7	230.1	240.1
Fluoride(1-1.5)	0.6	0.6	0.1
Iron(0.3)	0.3	0.3	0.3
Ammonia(1.5)	1.5	0.1	1.5
Arsenic(10ppb)	10ppb	Negative	10ppb

KAPOTA BALI I

Parameter	Summer	Monsoon	Winter
Date	02/06/2022	31/07/22	12/03/22
Time	12:00 am	1:00 pm	3:00 pm
Collected by	Dave, Marco	Umar, Charan	Abhishek
Location	Concrete cistern collecting runoff through pipes		
pH (6.5-8.5)	8	8.5	8.5
Dissolved Oxygen(5)	8	5	5
Turbidity(5-10)	<10	<6	<6
Nitrate(45Max)	10	<10	NVR
Phosphorous	0.1	1	1
Faecal Pollution(0)	Yes	Yes	Yes
Residual Chlorine(0.2)	<0.2	<0.2	<0.2
Chloride(250-1000)	177.25	63.7	180.4
Fluoride(1-1.5)	0.6	0.6	1.5
Iron(0.3)	<0.3	<0.3	<0.3
Ammonia(1.5)	1	1	1.5
Arsenic(10ppb)	Negative	50ppb	Negative

KAPOTA BALI II

Parameter	Summer	Monsoon	Winter
Date	02/06/2022	31/07/22	02/12/22
Time	10:00 am	1:00 pm	2:00pm
Collected by	Abhishek	Charan	Abhishek
Location	Roadside hand Pump from well		
pH(6.5-8.5)	8	8.5	8
Dissolved Oxygen(5)	6	5	5
Turbidity(5-10)	<10	6	7
Nitrate(45Max)	<1.0	<10	NVR
Phosphorous	No	No	No
Faecal Pollution (0)	No	Yes	Yes
Residual Chlorine(0.2)	<0.2	<0.2	<0.2
Chloride(250-1000)	212.7	70.8	68.8
Fluoride(1-1.5)	0.6	0.6	0.7
Iron(0.3)	0.2	0.2	0.1
Ammonia(1.5)	1	1	1
Arsenic(10ppb)	50ppb	50ppb	50ppb

RAKI

Parameter	Summer	Monsoon	Winter
Date	02/06/2022	28/07/22	25/02/22
Time	12:30 pm	1:00 pm	2:00 pm
Collected by	Lindsay, Andrea, Sonia	Abhishek	Charan and Saleem
Location	Tube well-comes from underground to spigot		
General Observations	Large garden, cows, goats		
pH(6.5-8.5)	8	8.5	8
Dissolved Oxygen(5)	3	5	5
Turbidity(5-10)	<10	<6	<6
Nitrate(45Max)	<10	<10	<10
Phosphorous	1	1	NVR
Faecal Pollution(0)	Yes, Fecal Contamination, E. Coli	No	No
Residual Chlorine(0.2)	<0.2	<0.2	NVR
Chloride (250-1000)	354.5	604.7	70.9
Fluoride(1-1.5)	0.6	0.6	0.6
Iron(0.3)	3	3	3
Ammonia(1.5)	1.5	1.5	1.5
Arsenic(10ppb)	Negative	Negative	Negative

GUJJAR COLONY

Parameter	Summer	Monsoon	Winter
Date	05/05/2022	07/06/22	10/03/22
Time	10:00 am	12:05 pm	1:00 pm
Collected by	Ghulam Hussein	Abhishek	Abhishek
Location	Water Tank in house		
Environmental Conditions	Sunny and windy		
General Observations	Bathroom tap and buffalo		
pH(6.5-8.5)	8	5	8
Dissolved Oxygen(5)	5.4	5.4	5.4
Nitrate(45Max)	10	40	10
Hardness(300-600)	136	400	400
Faecal Pollution(0)	No	No	Yes
Temperature (10-25deg.)	24	35	14
Residual Chlorine(0.2)	0	0	0
Chloride(250-1000)	38.9	300	35.4
Fluoride(l-1.5)	0.6	0.7	0.6
Iron(0.3)	0.3	0.3	0.3
Ammonia(1.5)	1	1	1
Arsenic (10ppb)	Negative	Negative	Negative

SATWARI

Parameter	Summer	Monsoon	Winter
Date	06/01/2022	01/06/22	01/03/22
Time	4:00 pm	5:00pm	6:00pm
Collected by	Dave	Abhishek	Abhishek, Dave
Location	Hand Pump		
Environmental Conditions	Hot, Sunny, Hazy		
General Observations	Fields surrounding village, a few Irrigated, near Border where signs were noted "Border Economic Development"		
Feces	Buffalo dung		
Animals	Buffalo, goats		
pH (6.5-8.5)	8	7	7
Dissolved Oxygen(5)	5	3	4
Nitrate(45Max)	40	42	42
Phosphorous	0.1	0.2	0.2
Faecal Pollution(0)	No	No	No

Parameter	Summer	Monsoon	Winter
Residual Chlorine(0.2)	<0.2	<0.2	<0.2
Chloride (250-1000)	70.9	70.9	700
Fluoride(1-1.5)	0.6	1.5	1.5
Ammonia(1.5)	<1.0	<1.0	<1.0
Arsenic(10ppb)	Positive	Positive	Positive

DAK BANGLA

Parameter	Summer	Monsoon	Winter
Date	06/01/2022	26/07/22	04/02/22
Time	10:00 am	11:00am	12:00pm
Collected by	Abhishek	Umar	Abhishek
Location	Handpump		
Environmental conditions	Hot, hazy, sunny		
General Observations	Irrigated fields, large herds, dung		
Litter	No	No	No
Feces	Yes, large piles		
Animals	Large hered of buffalo		
pH(6.5-8.5)	8	5	6
Dissolved Oxygen(5)	7	5	5
Nitrate(45max)	<10	<10	<10
Phosphorous	<0.1	<0.1	<0.2
Faecal pollution(0)	No	No	No
Residual chlorine(0.2)	<0.2	<0.2	<0.2
Chloride(250-1000)	88.625	880.3	881.2

Parameter	Summer	Monsoon	Winter
Fluoride(1-1.5)	0.6	0.6	1.5
Iron(0.3)	3	3	3
Ammonia(1.5)	<1.0	<1.5	<1.5
Arsenic(10ppb)	Negative	Negative	Negative

KACHA TALAB

Parameter	Summer	Monsoon	Winter
Date	06/06/2022	10/07/22	13/10/22
Time	11:00 am	12:00pm	1:00pm
Collected by	Perry	Abhishek	Umar
Location	From a house in front of the talab at Kacha Talab		
Environmental conditions	Sunny		
General Observations	Water is pumped up from a near by location		
pH(6.5-8.5)	8	5	6
Dissolved Oxygen(5)	Na	Na	Na
Turbidity(5-10)	Nil	Nil	Nil
Nitrate(45max)	<10	<10	<10
Phosphorous	0.1	0.1	0.1
Hardness(300-600)	n/a	n/a	n/a
Faecal pollution(0)	Yes		
Residual chlorine(0.2)	Nil	Nil	Nil
Chloride(250-1000)	71	71	73

Parameter	Summer	Monsoon	Winter
Fluoride(1-1.5)	0.6	0.4	0.1
Iron(0.3)	3	3	3
Ammonia(1.5)	1	1.5	1.5
Arsenic	Negative	Negative	Negative

BUDA AMARNATH

Parameter	Summer	Monsoon	Winter
Date	13/9/2022	14/06/22	13/03/22
Time	10:00 am	12:00 pm	1:00pm
Collected by	Sohan	Sohan	Umar
Location	Near Budha Amarnath Mandi, Mandi, Poonch		
Environmental conditions	Rainy		
General Observation	Covered cemented shelter near road		
pH(6.5-8.5)	7	7	7
Nitrate(45max)	9	9	9
Phosphorous	0.1	0.1	0.1
Faecal pollution(0)	Yes	No	No
Chloride(250-1000)	722	701	704
Fluoride(l-1.5)	0.6	0.6	0.6
Iron(0.3)	0.5	0.3	0.3
Ammonia(1.5)	2	1.5	1.5
Arsenic	Negative	Negative	Negative

BACBA WALI

Parameter	Summer	Monsoon	Winter
Date	12/9/2022	13/06/22	16/03/22
Time	10:00 am	11:00 am	12:00 pm
Collected by	Masood Ahmed	Abhishek	Abhishek
Location	Tanki water with attached bathroom in Kall, Poonch		
Environmental conditions	Sunny		
pH(6.5•8.5)	7	8	8.5
Nitrate(45max)	10	43	44
Phosphorous	0.2	0.2	0.2
Chloride(250-1000)	712	713	700
Fluoride(1-1.5)	0.6	0.3	0.4
Iron(0.3)	0.4	0.3	0.3
Ammonia(1.5)	1	1.5	1.5
Arsenic	10 ppb	Negative	Negative

CHASMA THANDA

Parameter	Summer	Monsoon	Winter
Date	12/9/2022	12/07/22	20/12/22
Time	3:10pm	4:00pm	6:00pm
Collected by	Abhishek	Sohan	Abhishek
Location	Kalai Poonch, Chasma Thanda near a Field		
Environmental conditions	Monson		
House hold effluents?	Yes		
Animals?	Yes		
Bank Stability/erosion?	Yes		
Agricultural runoff?	Yes		
Washing directly at the water?	Yes		
pH(6.5-8.5)	8	8.5	8.5
Turbidity(5-10)	50	10	10
Nitrate(45max)	45	45	45
Phosphorous	0.2	0.2	0.2
Faecal pollution(0)	Yes	Yes	Yes
Chloride(250-1000)	354	600	600

Parameter	Summer	Monsoon	Winter
Fluoride(1-1.5)	0.8	0.8	0.5
Iron(0.3)	0.5	0.5	0.3
Ammonia(1.5)	2	0.5	0.5
Arsenic	Negative	Negative	Negative

JAMIA MASJID

Parameter	Summer	Monsoon	Winter
Date	13/09/2022	14/06/22	18/10/22
Time	11:55am	12:55pm	6:09pm
Collected by	Irfan	Abhishek	Umar
Location	Jamia Masjid, Mohalla Chohana Khanater, Poonch, pipe line near Jamia Masjid		
Washing directly at the water?	Yes		
pH(6.5-8.5)	7	6	6
Nitrate(45max)	44	44	45
Phosphorous	0.1	0.1	0.1
Faecal pollution(0)	Yes	Yes	Yes
Fluoride(1-1.5)	0.6	1	1
Iron(0.3)	0.3	0.3	0.3
Ammonia(1.5)	1	2	0.1
Arsenic	Negative	Negative	Negative

KHANETER

Parameter	Summer	Monsoon	Winter
Date	13/09/2022	17/06/22	18/10/22
Time	10:00 am	12:00 pm	1:00 pm
Collected by	Irfan	Sohan	Abhishek
Location	Khaneter Water pump		
Environmental conditions	Sunny		
General observation	Water collection tanks dirty on inside		
Washing directly at the water?	Yes		
pH(6.5-8.5)	8	8	8.5
Nitrate(45max)	42	42	45
Phosphorous	0.1	0.1	0.1
Faecal pollution(0)	No	No	No
Chloride (250-1000)	71	700	701
Fluoride(1-1.5)	0.6	0.1	0.3

Parameter	Summer	Monsoon	Winter
Ammonia(1.5)	1	0.1	1.4
Arsenic	Na	Na	Na

MURI

Parameter	Summer	Monsoon	Winter
Date	13/9/2022	18/06/22	20/01/22
Time	10:00am	11:00am	12:00pm
Collected by	Farman Ali	Abhishek	Abhishek
Location	Mohallan Muri, Khaneter, Poonch		
General observation	Pipeline attached to Chashma		
Animals	Yes		
Washing directly at the water?	Yes		
pH(6.5-8.5)	8	8.5	8.5
Dissolved Oxygen(5)	50	51	51
Nitrate(45max)	44	44	44
Phosphorous	0.2	0.2	0.2
Faecal pollution(0)	Nil	Nil	Nil
Fluoride(1-1.5)	0.9	1.3	1.3
Iron(0.3)	0.7	0.2	0.2
Ammonia(1.5)	1.2	1.2	1.2
Arsenic	Positive	Positive	Positive

UDHARIAN

Parameter	Summer	Monsoon	Winter
Date	12/09/2022	27/07/22	31/03/22
Time	7:00pm	8:00pm	12:00pm
Collected by	Irfan	Abhishek	Ravi
Location	Mollah Udharain Kalai Poonch		
pH(6.5-8.5)	7	5	8.5
Nitrate(45Max)	40	40	42
Phosphorous	0.1	0.1	0.1
Hardness(300-600)	360	360	370
Faecal Pollution(0)	Yes	No	No
Fluoride(1-1.5)	0.6	1.5	1.5
Arsenic	Positive 10ppb	Positive 10ppb	Positive 10ppb
Ammonia(1.5)	1	1.5	1.5

KALAI

Parameter	Summer	Monsoon	Winter
Date	12/09/2022	04/06/22	08/02/22
Time	9:40am	12:00pm	1:00pm
Collected by	Farman	Abhishek	Abhishek
Location	Water tanki supply by PHE at H.S. Kalai Poonch		
General Observations	Tanki cemented and well covered		
pH(6.5-8.5)	8	8.5	8.5
Nitrate(45Max)	42	45	45
Phosphorous	0.1	0.2	0.1
Faecal Pollution(0)	Yes	Yes	Yes
Chloride(250-1000)	500	600	600
Fluoride(1-1.5)	1.3	1.2	1.3
Iron(0.3)	0.3	0.3	0.3
Ammonia(1.5)	1	1	1
Arsenic	Negative	Negative	Negative

CHOWADHI

Parameter	Summer	Monsoon	Winter
Date	13/09/2022	07/07/22	12/12/22
Time	10:00am	11:00am	12:00pm
Collected by	Abhishek	Abhishek	Ghulam Hussain
Location	Chowhana, Khanater, Poonch, pipeline From Chashma at Mohallah Chowan		
General Observations	Pipe line in field		
pH(6.5-8.5)	8	8.5	8.5
Nitrate(45Max)	35	35	35
Faecal Pollution(0)	Yes	No	No
Residual Chlorine(0.2)	0.2	0.2	0.2
Chloride(250-1000)	106	107	108
Iron(0.3)	Nil	Nil	Nil
Ammonia(1.5)	Nil	Nil	Nil
Arsenic	Negative	Negative	Negative

KURLIAN

Parameter	Summer	Monsoon	Winter
Date	14/03/2022	09/07/22	10/11/22
Time	10:00 am	12:00pm	1:00pm
Collected by	Abhishek	Abhishek	Sohan
Location	Spring (Water comes under hill) surrounded bywall (cemented), Kurlin Rajouri		
Environmental Conditions	Spring		
General Observations	Area of Bakerwal location (Near his homes), source underground water Depth:3 feet		
Litter	Animal dung and staw near by		
Animals	Drinking and Bathing		
Washing directly at the water	Yes		
pH(6.5-8.5)	8	8.5	8.5
Turbidity(5-10)	10	10	10
Nitrate(45Max)	29	29	29

Parameter	Summer	Monsoon	Winter
Hardness(300-600)	200	300	300
Chloride(250-100)	212.7	212	212
Ammonia(1.5)	Less than 0.1	Less than 0.1	Less than 0.1

GANDOH

Parameter	Summer	Monsoon	Winter
Date	26/03/2022	29/07/22	10/10/22
Time	10:00am	4:30pm	9:30am
Collected by	Mohd Riyaz	Sohan	Umar
Location	Gundoh Bawli		
Environmental Conditions	Clear Sky		
General Observations	In forest area, water flow near		
Animals	Drinking		
Water usage description:	Drinking and washing, mostly drinking		
pH(6.5-8.5)	8	8	8
Dissolved Oxygen(5)	5	5	5
Turbidity(5-10)	10	5	6
Nitrate(45Max)	40	40	40
Phosphorous	0.1	0.1	0.1
Hardness (300-600)	500	500	500
Faecal Pollution(0)	Yes	Yes	Yes

Parameter	Summer	Monsoon	Winter
Residual Chlorine(0.2)	0.2	0.2	0.2
Chloride (250-1000)	177.25	177	177
Fluoride(1-1.5)	Less than 0.6	Less than 0.6	Less than 0.6
Ammonia(1.5)	<1.0	<1.0	<1.0

MADRASA SANASAR

Parameter	Summer	Monsoon	Winter
Date	26/02/2022	01/06/22	14/11/22
Time	10:00 am	4:00pm	5:00pm
Collected by	Caltlin Collier	Abhishek	Ravi
Location	Spring just above the madrassa in Sanasar		
General Observations	Many stones around		
Litter	Fesmall bits of paper and plastic around. Animal dung present.		
Feces	Yes, animal		
Animals	Yes, drinking from stream, Buffalo and Goats mostly		
Agricultural runoff	None		
Washing directly at the water	Yes, washing clothes just below the main water collection site.		
Water usage description:	Drinking and washing		
pH (6.5-8.5)	9	8.5	8.5

Parameter	Summer	Monsoon	Winter
Dissolved Oxygen(5)	10	10	10
Turbidity(5-10)	10	6	6
Nitrate(45Max)	Less than 10	Less than 10	Less than 10
Phosphorous	Less than .1	Less than .1	Less than .1
Hardness(300-600)	160	170	170
Faecal Pollution(0)	Yes, Fecal Contamination, E.Coli	No	No
Chloride (250-1000)	35.4	35.4	35.4
Fluoride(1-1.5)	Less than 1	Less than 1	Less than 1
Iron(0.3)	Less than 0.3	Less than 0.3	Less than 0.3
Ammonia(1.5)	2.5	2.5	2.5

SANASAR

Parameter	Summer	Monsoon	Winter
Date	07/07/2022	04/07/22	10/12/22
Time	7:30pm	4:00pm	1:00pm
Collected by	Umar	Sohan	Ravi
Location	Upper Spring, Ahlbrechit house, Sanasar		
Environmental Conditions	Warm, Clearday		
General Observations	Covered with thorn bushes		
Feces	Yes		
Animals	Yes		
Riverbank vegetation	Grass, small shrubs		
Water usage description:	Cooking, drinking, cleaning		
pH(6.5-8.5)	8.5	8.5	8.5
Dissolved Oxygen(5)	3	3	3
Turbidity(5-10)	10	10	10
Nitrate(45Max)	Less than 10	Less than 10	Less than 10

Parameter	Summer	Monsoon	Winter
Phosphorous	Less than 0.1	Less than 0.1	Less than 0.1
Hardness (300-600)	100	200	100
Residual Chlorine(0.2)	Less than 0.2	Less than 0.2	Less than 0.2
Chloride (250-1000)	319	400	400
Fluoride(1-1.5)	0.6	0.6	0.6
Iron(0.3)	0.3	0.4	0.3
Ammonia(1.5)	2.5	2.5	2.5
Arsenic	Negative	Negative	Negative

BHADERWAH CBAKKA

Parameter	Summer	Monsoon	Winter
Date	02/07/2022	04/06/22	27/12/22
Time	10:00am	12:00pm	1:00pm
Collected by	Ravi	Umar	Sohan
Location	Lower Spring		
General Observations	Covered with thorn bushes to keep animals out, But animals still drink here.		
Feces	Yes		
House hold effluents	Laundry soap runs off below the spring		
Animals	Yes: buffalo, goats, horses, chicken		
Bank stability/erosion	Some erosion around bank from animal traffic		
Washing directly at the water	Yes, clothes around bank from animal traffic		

Parameter	Summer	Monsoon	Winter
Water usage description:	Bathing, washing clothes and dishes		
pH(6.5-8.5)	8.5	6.5	6.5
Dissolved oxygen(5)	6	5	5
Turbidity(5-10)	10	6	6
Nitrate(45Max)	Less than 10	<10	<10
Phosphorous	0.3	0.3	0.3
Hardness (300-600)	120	400	400
Residual Chlorine(0.2)	Less than 0.2	<2	<2
Chloride (250-1000)	450	450	450
Fluoride(1-1.5)	0.6	0.5	1.5
Iron (0.)	3	3	3
Ammonia(1.5)	2.5	2.5	2.5

BHALESA

Parameter	Summer	Monsoon	Winter
Date	26/03/2022	26/07/22	26/12/22
Time	10:00am	11:00am	12:00am
Collected by	Irfan Ali	Abhishek	Abhishek
Location	Bhalesa Spring		
General Observations	Little bit mud ner source and man go and banyan trees and some stones		
Bank Stability/erosion	Yes		
Water usage description:	Drinking and washing.		
pH(6.5-8.5)	7	7.5	7.5
Turbidity(5-10)	<10	<9	<9
Nitrate(45Max)	43.5	43.5	44
Phosphorous	<1.0	<1.0	<1.0
Hardness(300-600)	400	400	400
Faecal Pollution(0)	Yes	No	No

Parameter	Summer	Monsoon	Winter
Residual Chlorine(0.2)	<1.0	<1.0	<1.0
Chloride(250-1000)	177.25	177	177
Fluoride(1-1.5)	<0.6	<0.6	<0.6
Ammonia(1.5)	<1.0	<1.0	<1.0

The findings from our research on water contamination and scarcity in Jammu paint a concerning picture of the region's water quality. The analysis of sixty water samples revealed alarming results, with forty six samples showing fecal contamination, twenty containing arsenic, twelve exhibiting high levels of ammonia, and eleven having elevated iron content. Additionally, six samples were found to contain either nitrate, fluoride, or chloride. These results underscore the urgent need for comprehensive and immediate action to address the water crisis in Jammu. The presence of fecal contamination highlights the risks to public health, emphasizing the necessity of clean water for basic human needs. Moreover, the presence of toxic substances such as arsenic poses long-term health risks, making it imperative to implement rigorous filtration and treatment processes. The elevated levels of ammonia, iron, nitrate, fluoride, and chloride further emphasize the complexity of the contamination issue, requiring multifaceted solutions. It is crucial for policymakers, communities, and environmental agencies to collaborate on implementing stringent water quality standards, improving sanitation infrastructure, and promoting public awareness. By prioritizing these measures, we can work towards ensuring access to safe and clean water for the residents of Jammu, thus safeguarding public health and enhancing overall quality of life.

Conclusion

In the heart of Jammu, where the picturesque landscapes and vibrant culture paint a serene picture, lies a crisis that threatens the very essence of life: water scarcity and contamination. Through the pages of this book, we have delved deep into the intricate web of challenges faced by the people of Jammu, as they grapple with the diminishing water sources and the perils of contamination.

Our exploration began by unraveling the historical roots of water scarcity, tracing its evolution from a localized concern to a pressing regional issue. We witnessed the exponential growth of population and urbanization, coupled with the effects of climate change, exacerbating the scarcity of this precious resource. Simultaneously, the specter of contamination loomed large, infiltrating the very lifelines that the people depended upon for survival.

The alarming revelations presented in the preceding chapters serve as a clarion call for immediate action. We have seen the toll that contaminated water takes on the health of individuals, leading to a myriad of diseases and a burden on the already stretched healthcare system. The socio-economic fabric of communities is strained, as families struggle to secure clean and safe water for their daily needs.

One of the key takeaways from this exploration is the vital role of awareness and advocacy. Empowering the community with knowledge about water conservation, purification techniques, and the significance of sustainable practices can pave the way for change. Moreover, advocating for policy reforms, infrastructure development, and stringent regulations are essential steps in the battle against water scarcity and contamination.

Addressing the water crisis in Jammu necessitates collaborative efforts from all stakeholders. Government bodies, NGOs, local communities, and international organizations must join hands in a united front. Community-led initiatives, harnessing traditional wisdom and modern technology, can create innovative solutions tailored to the specific needs of the region. Education, awareness campaigns, and capacity building should be prioritized to ensure a sustainable future for generations to come.

In the face of adversity, there is always room for hope. The resilience of the people of Jammu, their determination to overcome challenges, and the collective will to safeguard their most precious resource are beacons of hope. As we conclude this journey, let us carry forward the knowledge gleaned from these pages and transform it into action.

This book is not an endpoint but a beginning. It is a catalyst for change, inspiring individuals and communities to take decisive steps toward a water-secure future. As we turn the final page, let us embark on a new chapter—one where every drop of water is cherished, where contamination is eradicated, and where the beauty of Jammu's landscapes is mirrored in the purity of its water sources.

In the words of an ancient proverb, "Thousands have lived without love, not one without water." Let us ensure that the people of Jammu, and indeed all of humanity, do not have to live without this fundamental essence of life. Together, let us script a future where water scarcity and contamination are mere chapters in history, and the legacy we leave behind is one of abundance, health, and harmony.